Black Hole Orbits

The Relativistic Effects on a Spacecraft Trajectory Around a Rotating Blackhole

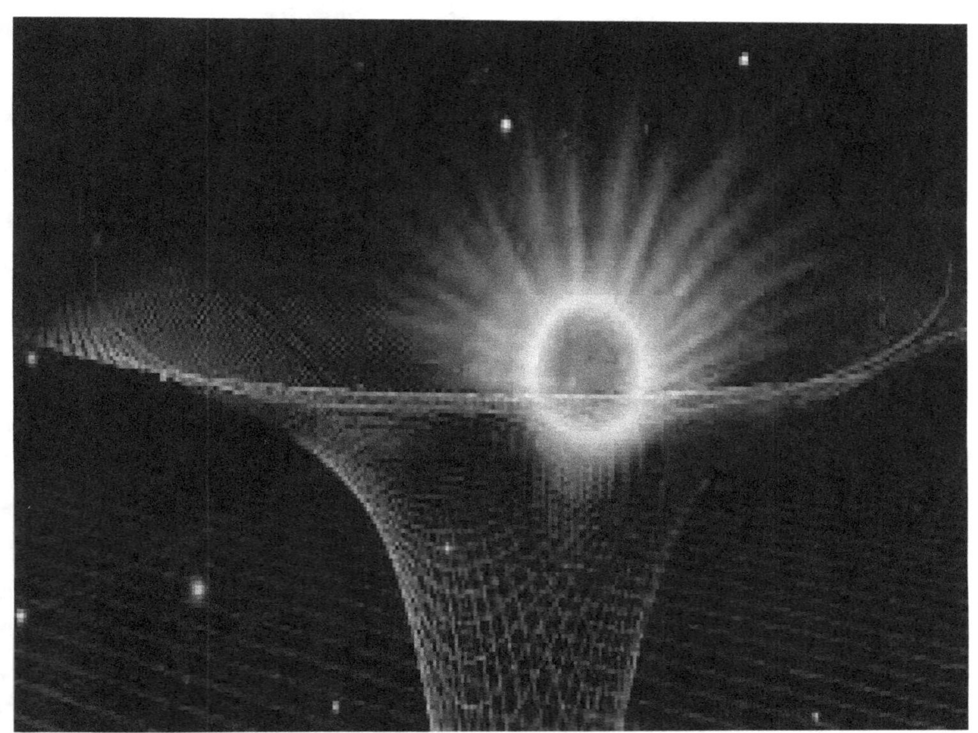

Dr Ugur GUVEN

Guven Publications

Disclaimer

Although the author and publisher have made every effort to ensure that the information in this book was correct at press time, the author and publisher do not assume and hereby disclaim any liability to any party for any loss, damage, or disruption caused by errors or omissions, whether such errors or omissions result from negligence, accident, or any other cause.

Copyright

Black Hole Orbits: The Relativistic Effects on a Spacecraft Trajectory Around a Rotating Blackhole

© April 2022, by Dr. Ugur GUVEN
Self-published through Amazon under Guven Publications
ISBN: 9798811209996

Published in London
Contact: drguven@live.com

All rights reserved.
No part of this publication may be reproduced, stored in a retrieval system, stored in a database and / or published in any form or by any means, electronic, mechanical, photocopying, recording or otherwise, without the prior written permission of the publisher.

ACKNOWLEDGEMENT

The author thanks his students who have been instrumental in supporting this research and this book.

NOMENCLATURE

- SR – Special Relativity

- KBH – Kerr Black Hole

- K-Carter's Constant

- E-Energy of orbiting particle or test particle

- a= angular momentum of Black Hole per unit Mass

- L=angular momentum of test particle

- l= angular momentum of test particle per unit rest mass

- $\varepsilon =$ measure of Killing Metric which is 1 for test particle at rest

ABSTRACT

1972, the year responsible for the study of black holes, when a team of great astrophysicist brains came together to analyze radiations ranging from radio waves to X-rays and discovered one of the most magnificent bodies of the cosmos, the black hole. Cygnus X-1 is the name it was given, a black hole 14,000 light years away from our home. Ever since then, research has been going on to study the nature of the black hole which gets more mysterious with every trait it reveals.

The Black Hole is born when a massive star implodes due to its own gravity. The horizon of the black hole is a very special region which is the critical limit to observe and study the black hole. Beyond or inside the horizon, everything is lost and absorbed into the black hole. Therefore, by abiding to certain constraints at the horizon we can experience what happens at the horizon and understand the black hole.

Moreover, on the basis of mass the black holes may be distinguished as stellar and galactic. A stellar black hole is one weighing a few solar masses and galactic is one weighing millions of solar masses. It is interesting to note that larger the black hole, lesser will be the tidal effects at the event horizon. Event horizon varies in a rotating and a non-rotating black hole. With a non-rotating black hole, the event horizon is spherical and there is zero angular momentum.

To study the effects of the vicinity of the event horizon on the orbit of a spacecraft is an essential and interesting task as a step towards the future space technology. The first half of the project will involve study on a test mass in proximity of the event horizon followed by extending the concepts on a whole spacecraft which will involve tidal forces. In our study three concepts: a) gravitational potential (GM/R) b) gravitational forces (GM/R^2) and c) tidal forces (GM/R^3) will be used.

This book will enable astronauts, scientists and engineers to design future KBH missions in keeping with the time warps that are caused around the black hole. This book is especially recommended for undergraduate students in Astronomy and Aerospace Engineering Fields.

TABLE OF CONTENTS

LIST OF FIGURES

CHAPTER 1: INTRODUCTION

1.0 BLACK HOLE:

A Black hole is, by definition, a region in space-time in which the gravitational field is so strong that it precludes everything, even light, from escaping to infinity.

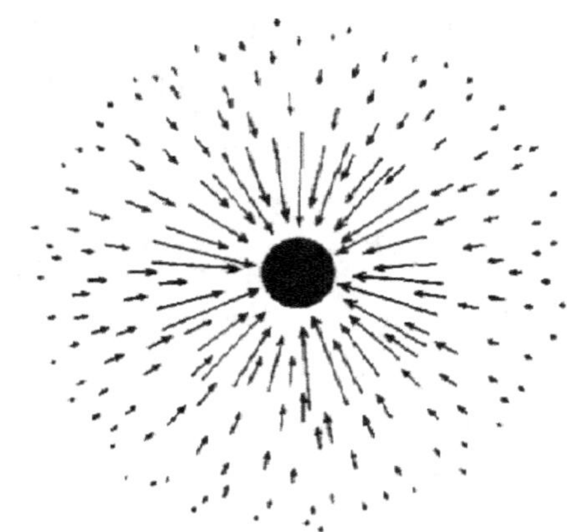

Figure 1: Forces around a Black Hole

1.1 HISTORY

1972, the year responsible for the study of black holes, when a team of great astrophysicist brains came together to analyze radiations ranging from radio waves to X-rays and discovered one of the most magnificent and strange body of the cosmos, the Black Hole. Cygnus X-1 is the name it was given, a black hole 14,000 light years away from our home planet. Ever since then, research has been going on to study the nature of the black hole which gets more mysterious with every study.

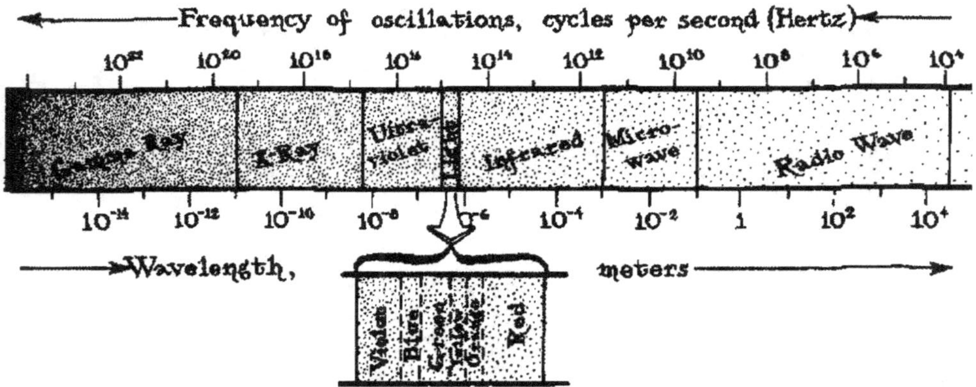

Figure 2: Band of light waves starting from strongest to weakest waves

Degraded remnants of light waves accompanied with a late phase of an initial explosion are a indication of the possible existence of a black hole.

1.2 SPACETIME CURVATURE OF BLACK HOLE AND ITS RADIUS

The Black bends the Space-time grid around itself. The vertical lines of the grid depict the line of time coordinates and the horizontal lines depict the dimension of space. The deep as infinite bend created due to a Black Hole
Drastically bends this space-time continuum and brings about extreme dilation in time and space, dilations to an extent that may even stop time or maybe take you back in time.

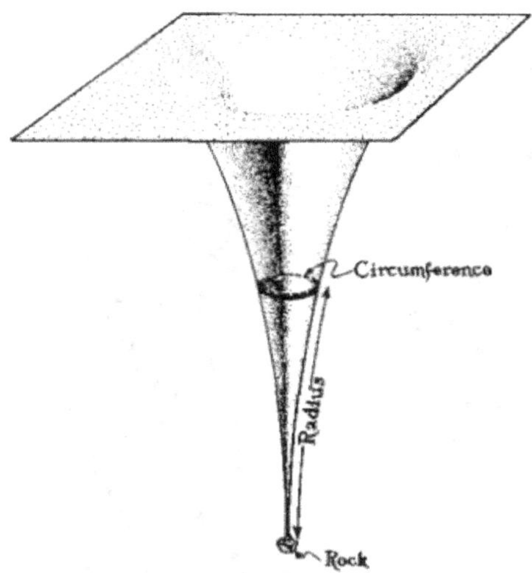

Figure 3: True radius of a Black Hole

The radius of a Black Hole is not what we obtain by simply calculating velocity around its event horizon and solving it for radius using $vt = 2\pi r$. The "radius of event horizon " is measure of it location assuming Black Hole to be a sphere. However the radius of a Black Hole s defined based on the bend it creates in spacetime curvature. The illustration is as shown above in the diagram.

1.3 GRAVITATIONAL REDSHIFT

This process is popularly known as the Einstein shift also.. When radiations from a source in a given magnetic field is observed from a gravitational field which is at a relatively higher gravitational potential , the electromagnetic radiation tend to get shifted. They eventually reduce in frequency, i.e. they get red shifted. This is a consequence of gravitational time dilation. This phenomenon states that -if one is outside of an isolated gravitational source, the rate at which time passes increases as one moves away from that source. As frequency is inverse of time (specifically, time required for completing one wave oscillation), frequency of the electromagnetic radiation is reduced in an area of higher gravitational potential.

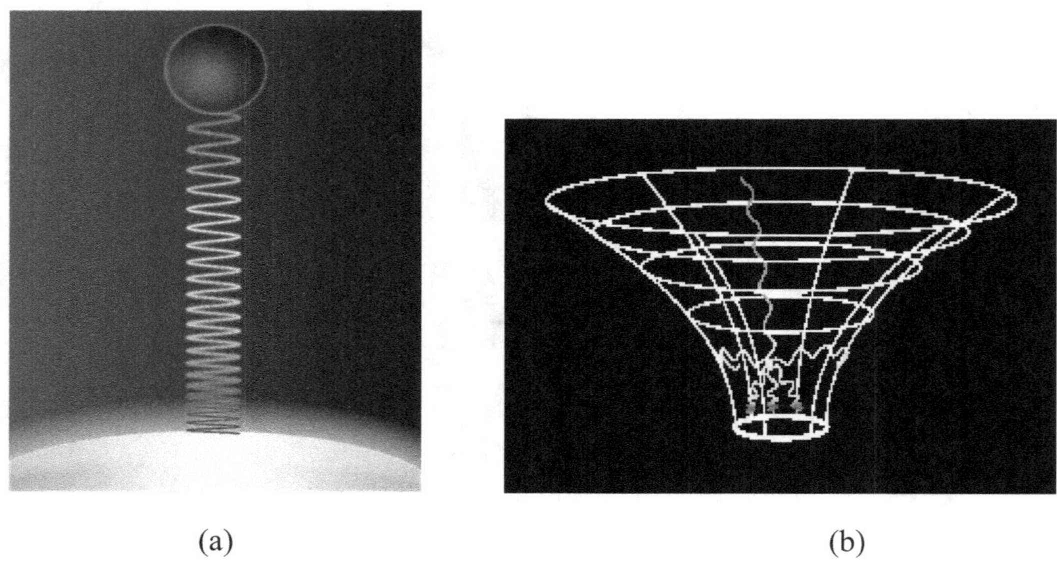

(a)   (b)

Figure 4: (a) Redshift explained generally; 10 (b) Redshift along space curve

1.4 APPROACH

The event horizon is an absolutely sharp edge of "no escape" just like the horizon on earth beyond which nothing can be seen. For a Kerr Black Hole this event horizon is a ring and it is so because of the additional factor of angular momentum. We have chosen a Kerr black hole because it is easier to hover over the horizon due to less tidal forces.

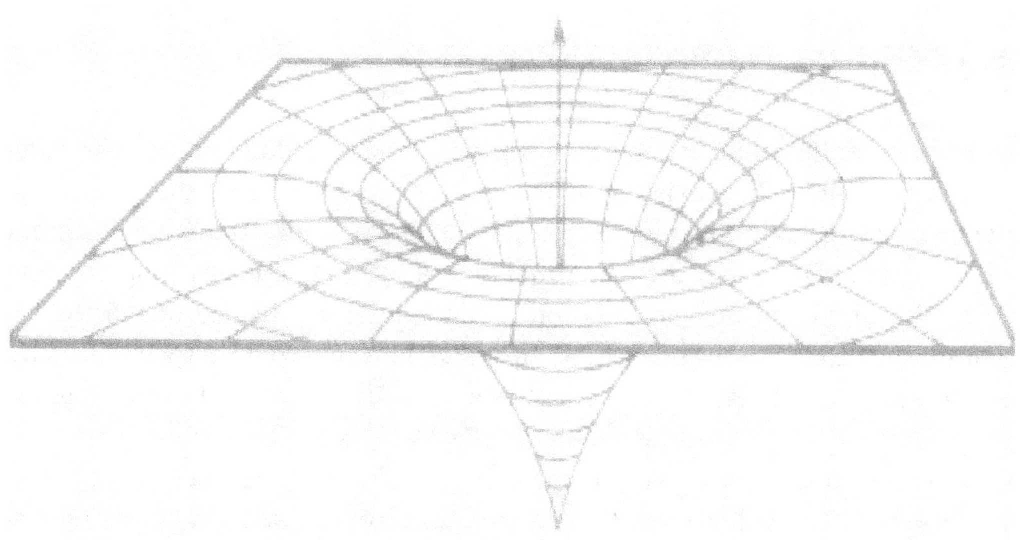

Figure 5: The vertical and horizontal lines on spacetime continuum

The objective of our team is to study the gravitational behaviours close to the event horizon in a black hole. By using concepts of orbital mechanics, Newtonian physics and relativity we will study the motion of a test point mass in proximity of the horizon and verify the results by software simulation.

CHAPTER 2: LITERATURE REVIEW

Black holes are the strangest objects in the Universe. Although black holes cannot be seen, we know they exist from the way they affect nearby dust, stars and galaxies. Most galaxies, including the Milky Way, have supermassive black holes at their centres. Do not let the name fool you: a black hole is anything but empty space. Will an astronaut who falls into a black hole be crushed or burned? The death of star explosion scatters most of a star into the void of space but leaves Black holes. Black holes are objects so dense, and with so much mass, that even light cannot escape it. Most people think of a black hole as a voracious whirlpool in space, sucking down everything around it, which in itself is a serious misconception. Black holes are formed from the cores of super massive stars and can best be described as regions of space where so much mass is concentrated that nothing can escape it gravitational pull. Black holes are places where ordinary gravity has become very extreme. Once inside, nothing can escape a black hole's gravity — not even light.

The idea of the existence of the black hole may seem too fictitious but the beauty is that it is real. No human from Earth has ever been too close to the black hole to be able to understand its behaviour and report the experiences to Earth. It is beyond our technological scope as even the nearest black hole Hades is light years away. With the current technology, the scientists on Earth will have to go into hibernation to wake up after over a hundred years and witness the results that human astronaut send to Earth. Researchers and experts have only been able to predict as to what will happen if an object or an astronaut was to approach the vicinity of the black hole. As we already know black hole is simply a huge body with a gravitational pull so high that even light itself cannot escape it. Moreover, it is born when a massive star is about to die, i.e. when it implodes due to its own gravity

Tensor Calculus, Concept of Special Relativity, Lorentz Transformations, Kerr Solutions, Frame Dragging, gravitomagnetic moment, values of conserved and non-con served quantities such as K, E, a, J, L etc. governing the motion around the Black Hole have been used to estimate the orbit around the Black hole. Our coding has been done in JAVA. The program essentially accepts limited values, pertaining to which it derives other necessary values based on the background equation provided to it. Based on these calculations, we get the orbit t

CHAPTER 3: METHODOLOGY

3.0 ACTION PLAN TIMELINE

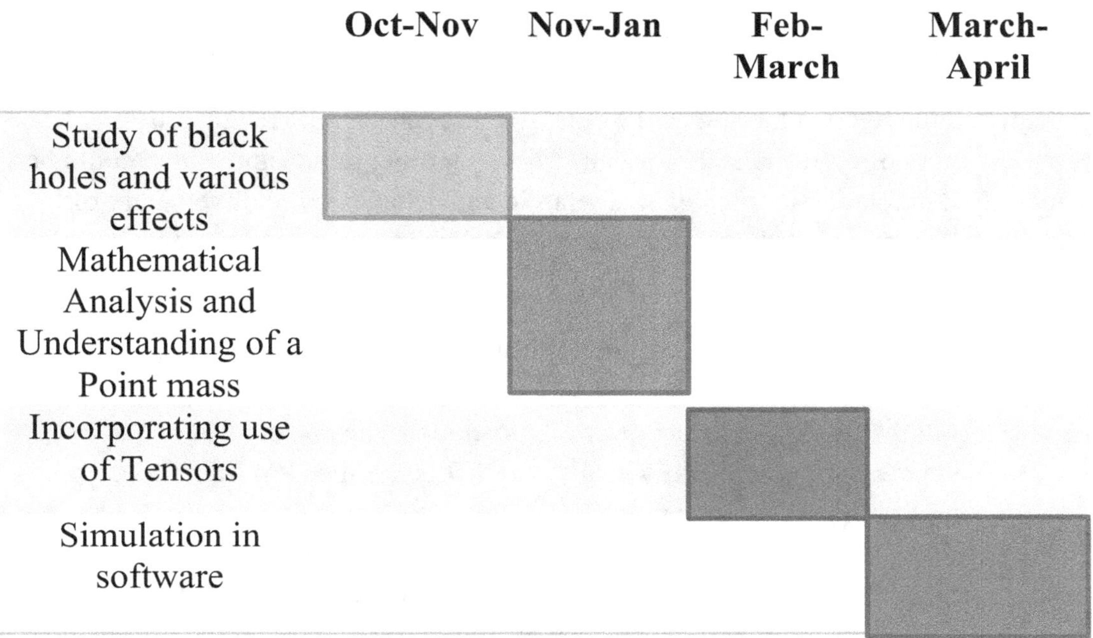

Figure 6: Gantt Chart

3.1 ORBIT MATHEMATICS

3.1.0 Einstein Special Relativity of Space-Time

"Your space is a mixture of my space and my time and my space is a mixture of your space and your time "

– Einstein

Einstein destroyed the idea of absolute space, he said that length, breadth and height are variable and depend on the relative motion of the observer measuring them and the relative motion of the objects being measured. He ignored absoluteness of time and said it was just like any other dimension which could be measured in relative terms, i.e., it is relative and depends on relative nature of events of observer. With his ignorance of absoluteness, Einstein came up with his postulates of relativity:

    i.   *Absoluteness of speed of light*: The space and time may vary, may be relative yet they must be framed in a manner that they should make speed of light absolutely the same in all directions. This means that one and only one measure is static universally and that is the speed of light. Speed of light is an invariant tensor.

    ii.   *Postulate of relativity*: The laws of physics are applied unbiased to all kinds of motion in the universe. Whatever be the situation or world we are in, the laws of physics govern all dynamics. No motion must be defined as the absolute motion.

3.1.1 Event Space and Light Cone

An event space is an R^4(four dimensional real plane) whose points are events, coordinated by $(x^i) = (x^0, x^1, x^2, x^3)$, where $x^0 = ct$ is a temporal coordinate (time dependent) and the rest three are the rectangular positional coordinates, of a given event.

A flash of light at (0,0,0,0) send a out an expanding spherical equation:

$$x^2 + y^2 + z^2 = c^2 t^2 \text{ or } (x^0)^2 - (x^1)^2 - (x^2)^2 - (x^3)^2 = 0 \qquad (3.1)$$

This equation represents the equation of the light cone in the event space relative to an inertial plane represented by x^i.

The Figure below gives the projection of the light cone onto the hyperplane $x^3 = 0$.

Note: The projection of the light cone equation will remain the same in any other inertial plane. This is because velocity of light is an invariant tensor, same for all observers.

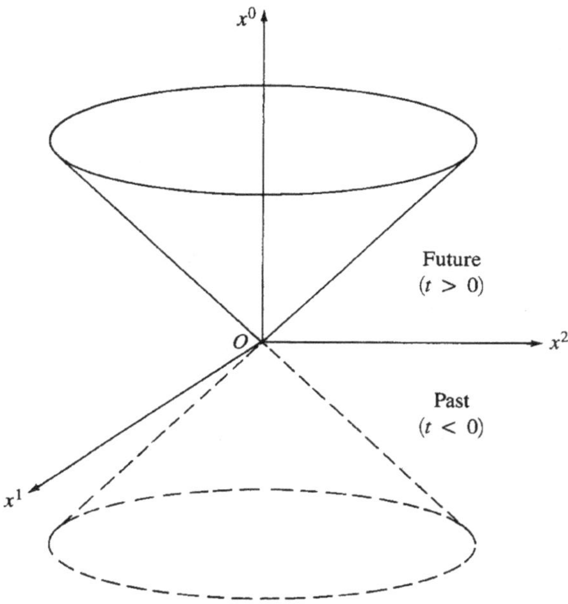

Figure 7: Projection of the Light Cone

3.1.2 Lorentz Matrix and transformation

Invariance of light cone equation can be given as
$$g_{ij}x^i x^j = 0 = g_{ij}\bar{x}^i \bar{x}^j \tag{3.2}$$
where $g_{00} = 1, g_{11} = g_{22} = g_{33} = -1$ and $g_{ij} = 0$ for $i \neq j$.

Substituting T: $\bar{x}^i = a_j{}^i x^i$ in Eq. (#), we get
$$g_{ij}a_r{}^i a_s{}^j = g_{rs} \tag{3.3}$$

Simple Lorentz Matrix

$$\mathcal{T} : \begin{cases} \bar{x}^0 = a_0^0 x^0 + a_1^0 x^1 \equiv ax^0 + bx^1 \\ \bar{x}^1 = a_0^1 x^0 + a_1^1 x^1 \equiv dx^0 + ex^1 \\ \bar{x}^2 = x^2 \\ \bar{x}^3 = x^3 \end{cases} \tag{3.4}$$

Where
$$a^2 - d^2 = 1\,, b^2 - e^2 = 1\ \ and\ \ ab - de = 1 \tag{3.5}$$

$$d = -\left(\frac{v}{c}\right)a - \equiv \beta a \quad and \quad a = e$$

$$(3.6)$$

By SR $\beta = \frac{v}{c}$ and $a = 0$ (since the two clocks in off-orbit and in-orbit are taken to be running in the same sense)

$$a = (1 - \beta^2)^{-1/2} = e \qquad b = -\beta(1 - \beta^2)^{-1/2} = d$$

$$(3.7)$$

The coordinate transformation in the simplest form can be written as:

$$\bar{x}^0 = \frac{x^0 - \beta x^1}{\sqrt{(1 - \beta^2)}}$$

$$\bar{x}^1 = \frac{-x^1 + \beta x^0}{\sqrt{(1 - \beta^2)}}$$

$$\bar{x}^2 = x^2$$

$$\bar{x}^2 = x^2$$

Or

$$A = \begin{pmatrix} \frac{1}{\sqrt{1-\beta^2}} & \frac{1}{\sqrt{1-\beta^2}} & 0 & 0 \\ \frac{1}{\sqrt{1-\beta^2}} & \frac{1}{\sqrt{1-\beta^2}} & 0 & 0 \\ 0 & 0 & 1 & 0 \\ 0 & 0 & 0 & 1 \end{pmatrix}$$

$$(3.8)$$

Above matrix is the simple transformation matrix by Lorentz.

3.1.3 Boyer Lindquist Coordinates

For an understanding of above formulas in 3D space, the Boyer Lindquist coordinates may be converted as :

$$x = \sqrt{(r^2 + a^2} \sin\theta \cos\varphi$$

$$y = \sqrt{(r^2 + a^2} \sin\theta \sin\varphi$$

$$z = r\cos\theta$$

$$(3.9)$$

3.1.4 The Kerr Metric

The orbit around a rotating Black Hole depends on three parameters:

1. Mass
2. Charge (Taken Zero)
3. Angular Momentum

Particularly known as three degrees of freedom on which the orbit parameters depend.

For a Schwarzschild metric which was formulated in 1916 by Schwarzschild was for the stationary black holes wherein the deciding parameter was only Mass, i.e. one degree of freedom. However, a rotating black hole also curves the space-time around itself because of its angular momentum. How this works? Rotational energy is a direct implication of the angular momentum which is thereby transferred to the surrounding space, surrounding gravitational grid and thus forms a source of rotating field by combining with the gravitational field.

In 1968, Astrophysicist Roy Kerr developed a metric for the solution of rotating black holes. It is popularly known as the Kerr metric. The Kerr metric is not based on three dimensional space. It is a formulation in four dimensional space using the Boyer-Lindquist coordinates. These coordinates are (t, r,θ,φ). The Kerr Metric given by Roy Kerr is dependent of Mass and Angular momentum of the rotating black hole.

The Kerr metric is given by:

$$ds^2 = -\left(1 - \frac{2Mr}{\rho^2}\right)dt^2 - \frac{4Mra\sin^2\theta}{\rho^2}d\varphi dt + \frac{\rho^2}{\Delta}dr^2 + \rho^2 d\theta^2 +$$
$$\left(r^2 + a^2 + \frac{2Mra^2\sin^2\theta}{\rho^2}\right)\sin^2\theta d\varphi^2 \tag{3.10}$$

In the above metric solution,

1. 'a' is known as the Kerr parameter and is the Angular Momentum of the black hole per unit mass of the black hole. $a = \frac{J}{M}$. J is the Angular Momentum of the KBH.
2. $\rho^2 = r^2 + a^2\cos^2\theta$
3. $\Delta = r^2 - 2Mr + a^2$
4. r is the radial distance from the KBH.

Important points to note:

- Since we have assumed zero charge on the Black Hole. Q value is zero.
- The above solution is a vacuum solution, i.e. it is valid only in vacuum (presence of no matter).

- To obtain the Schwarzschild solution directly from the above metric, we can substitute $a = 0$.
- At large distances, the curvature vanishes. For $r >> M$ and $r >> a$.
- The above equation is not dependent on the coordinates t and φ.

The above result can be used to obtain even the boundary of the event horizon.

When $\Delta \to 0$ at $r_{\pm} = M \pm \sqrt{(M^2 - a^2}$ $\qquad$ (3.11)

At r_{+}, we get the event horizon.

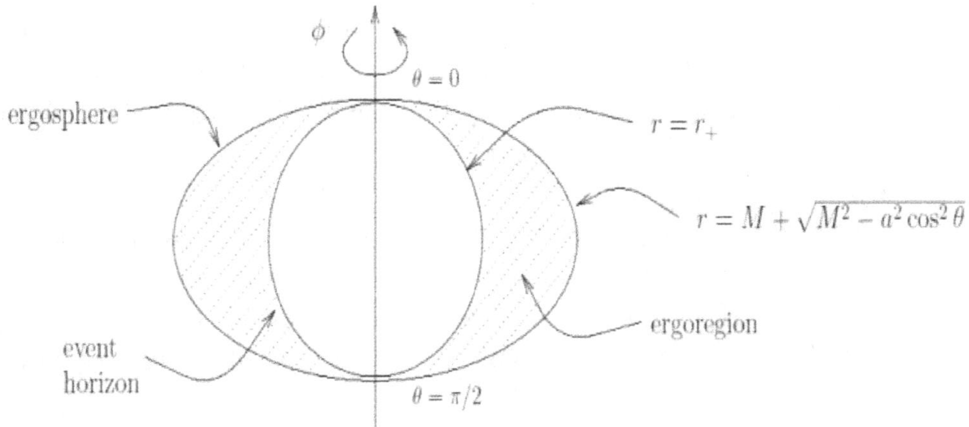

Figure 8: Ergosphere around a KBH

For obtaining the singularity, naturally the space $ds \to \infty$ and $\rho \to 0$
Interestingly the extreme Black Holes may be obtained by setting $a = M$. This gives us $J_{max} = M^2$.

3.1.5 Condition on Effective Potential

The standard equation for derivative of R with respect to proper time τ is given as

$$R^4 (\frac{dR}{d\tau})^2 = \left(R^2 \tilde{S}^2 + 2R\tilde{S}^2 + R^4\right)\tilde{E}^2 - \left(4R\tilde{S}\tilde{L}\right)\tilde{E}$$
$$- \left(\tilde{L}^2 R^2 - 2\tilde{L}^2 R + R^2(R^2 - 2R + \tilde{S}^2)\right)$$

$$(3.12)$$

$\tilde{S} = |J|/M^2$, KBH Angular Momentum per square mass.

$R = r/M$, radial distance from KBH divided by Mass of KBH

$\tilde{L}$ is the orbital angular momentum of the particle in orbit normalized with respect to its mass, m, and the KBH mass, M.
Equation (#0) corresponds to:

$$(\tilde{E} - \tilde{V}_+)(\tilde{E} - \tilde{V}_-) = 0$$

$$(3.13)$$

$$\tilde{V} \text{ is th potential,} \qquad \tilde{E} \text{ is the energy of the orbital element}$$

Solving above equation gives

$$\tilde{V}_\pm = \frac{-b \pm \sqrt{b^2 - 4ac}}{2a}$$

$$(3.14)$$

$$a = -(R^2\tilde{S}^2 + 2R\tilde{S}^2 + R^4)$$

$$b = 4R\tilde{S}\tilde{L}$$

$$c = \left(\tilde{L}^2 R^2 - 2\tilde{L}^2 R + R^2(R^2 - 2R + \tilde{S}^2)\right)$$

The equations for the radius of a circular or elliptical LSO can be calculated through a mathematical treatment of the following two equations:

$$\frac{d\widetilde{V_+}}{dR} = 0$$

$$(3.15)$$

$$\frac{d^2\tilde{V}_+}{dR^2} \leq 0$$

$$(3.16)$$

3.1.6 Orbital Angular Momentum ($\tilde{L}$)
Equating derivative of $\tilde{V}_+$ with respect to R to zero, we will obtain a quadratic equation which will give us the value of $\tilde{L}$. By doing so, the simplified quadratic equation obtained is :

$$R^3\left(9R - 6R^2 + R^3 - 4\tilde{S}^2\right)\tilde{L}^4 - 2R^2\left(-3R^4 + R^5 - 12\tilde{S}^2 R + 6R^2\tilde{S}^2 + 2R^3\tilde{S}^2 + 5\tilde{S}^4 + \tilde{S}^4 R\right)\tilde{L}^2 + (R^2 + 2R^2\tilde{S}^2 - 4\tilde{S}^2 R + \tilde{S}^4)^2 = 0 \qquad (3.17)$$

Solving above quadratic equation, we get two solutions for $\tilde{L}$, i.e., one for prograde orbit and the other for the retrograde orbit.

Prograde Orbit:

$$\tilde{L}^2{}_{Pro} = (-3R^6 + R^7 - 12R^3\tilde{S}^2 + 2R^5\tilde{S}^2 + 5\tilde{S}^4R^2 + \tilde{S}^4R^3 - 2\tilde{S}(3R^2 + \tilde{S}^2)(R^2 - 2R + \tilde{S}^2)\sqrt{R^3})(R^3(9R - 6R^2 + R^3 - 4\tilde{S}^2))^{-1}$$

$$(3.18)$$

Retrograde Orbit:

$$\tilde{L}^2{}_{Retro} = \left(-3R^6 + R^7 - 12R^3\tilde{S}^2 + 6R^4\tilde{S}^2 + 2R^5\tilde{S}^2 + 5\tilde{S}^4R^2 + \tilde{S}^4R^3 \right.$$
$$- 2\tilde{S}(3R^2 + \tilde{S}^2)(R^2 - 2R$$
$$\left. + \tilde{S}^2)\sqrt{R^3}\right)\left(R^3(9R - 6R^2 + R^3 - 4\tilde{S}^2)\right)^{-1}$$

$$(3.19)$$

3.1.7 Radii For Prograde and Retrograde

From equation 3.17, we can infer that derivative of $\tilde{L}^2$ with respect to R must be zero.

Putting $\frac{d(\tilde{L}^2)}{dR}=0$,

We get the simplified result as:

$$(9\tilde{S}^4 - 28\tilde{S}^2R - 6\tilde{S}^2R^2 + 36R^2 - 12R^3 + R^4) = 0 \qquad (3.20)$$

Above equation is a quartic equation. This equation is solved by converting to a companion matrix which is solved for eigenvalues which gives analytical solutions for both the prograde and retrograde orbits.

The solutions thus obtained are:

$$R_{pro} = 3 + \sqrt{Z} - \sqrt{\frac{16\tilde{S}^2}{\sqrt{Z}} - Z + 3(3 + \tilde{S}^2)}$$

$$(3.21)$$

$$R_{retro} = 3 + \sqrt{Z} + \sqrt{\frac{16\tilde{S}^2}{\sqrt{Z}} - Z + 3(3 + \tilde{S}^2)}$$

$$(3.22)$$

Where

$$Z = 3 + \tilde{S}^2 + \left(3 + \tilde{S}\right)\{(1 + \tilde{S})(1 - \tilde{S})^2\}^{1/3} + \left(3 - \tilde{S}\right)\{(1 - \tilde{S})(1 + \tilde{S})^2\}^{1/3} \quad (3.23)$$

3.1.8 Frame Dragging

The Black Hole curves space-time and it is capable of dragging the space around itself while rotating. Mathematically speaking, the Kerr metric contains non-diagonal components $g_{03} = g_{30}$.

Eventually, if a particle is dropped with an initial radial velocity onto a KBH, it will acquire perpendicular(non-radial) components of motion as it is under free fall in the gravitational field.

Following formula gives a measure of $\frac{d\varphi}{dt}$

$$\frac{d\varphi}{dt} = \frac{1}{\Delta}\left[\left(1 - \frac{2M}{r}\right)l + \frac{2Ma}{r}\varepsilon\right] \rightarrow \frac{d\varphi}{dt} = \frac{1}{\Delta}\left(\frac{2Ma}{r}\right) \qquad (3.24)$$

$$\frac{\varepsilon^2 - 1}{2} = \frac{1}{2}\left(\frac{dr}{dt}\right)^2 + V_{eff}(r, \varepsilon, l) \rightarrow \frac{dr}{dt} = \sqrt{\frac{2M}{r}\left(1 + \frac{a^2}{r^2}\right)} \qquad (3.25)$$

The above two equations may be combined to give the following result:

$$\frac{d\varphi}{dt} = \frac{\frac{d\varphi}{d\tau}}{\frac{dr}{d\tau}} = -\frac{2Ma}{r\Delta}\left[\frac{2M}{r}\left(1 + \frac{a^2}{r^2}\right)\right]^{-1/2}$$

$$(3.26)$$

According to the above Equation, the particle is dragged along angle φ as it falls radially toward the KBH, despite the fact that no forces really act on it along φ.

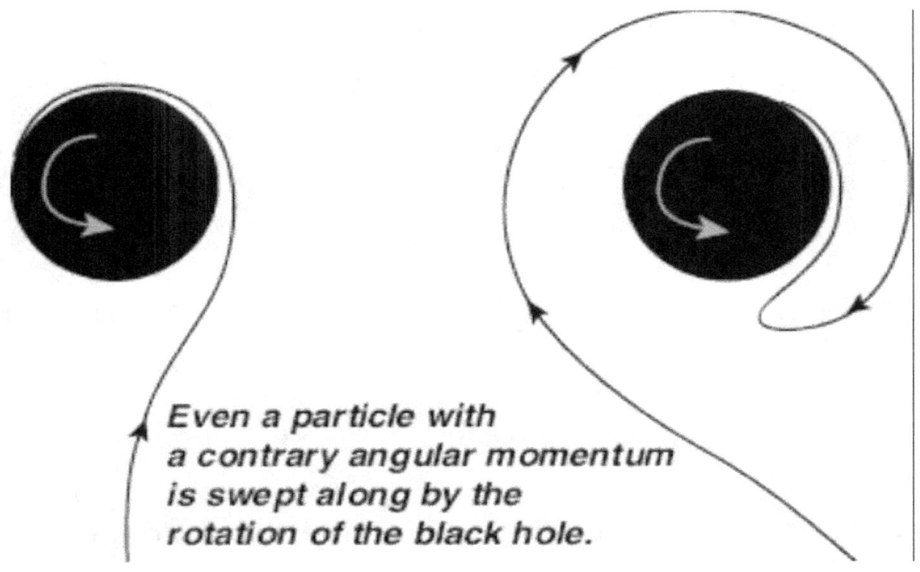

Figure 9: Effect of Frame Dragging

Our simulation has included the effect of frame dragging. Above figure shows the effect of frame dragging on a particle in orbit in as sense opposite to the rotation of the black hole. Frame Dragging essentially occurs in the region surrounding the event horizon, i.e., the ergosphere. The energy extractable from the black hole in the form spin energy is given as $0.29Mc^2$.

Frame dragging is also known as the "Dragging of Inertial Frames" and produces the precession in gyroscope. This effect of Precession is known as Lense Thirring effect.

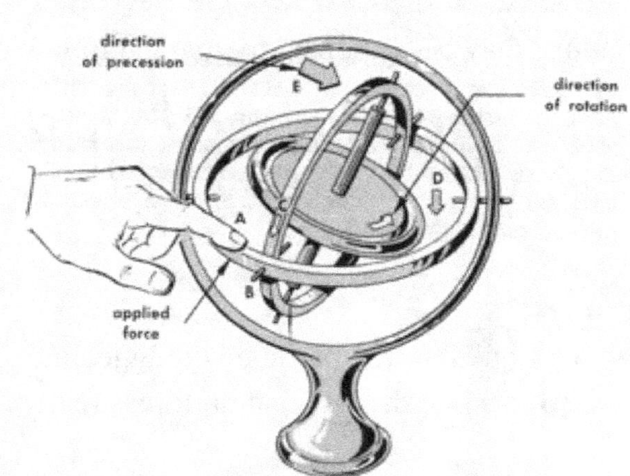

Figure 10: Precession in Gyroscope

3.1.9 Explanation for Effective Potential

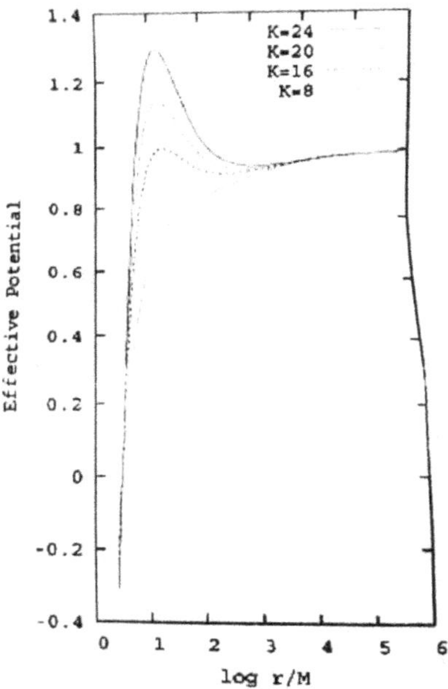

Figure 11: Effective potential v/s log r/M

$$V^2 = \Delta(K + r^2)/(r^2 + a^2)^2 \tag{3.27}$$

Value of K has been taken as 14.783. The Carter's constant is a unique value which remains conserved for the motion around the Black Hole. So for a test particle in orbit around a KBH, four quantities, i.e., energy, angular momentum, rest mass and K remain conserved.

$$K = C + (L - aE)^2 \tag{3.28}$$

$$where\ C = p_\theta{}^2 + cos^2\theta\{a^2(m^2 - E^2) + \left(\frac{L}{sin\theta}\right)^2\} \tag{3.29}$$

3.1.10 Various Parameters used in the software

1. v_{ring}: Initial four velocity of the orbiting particle.The four point velocity is the velocity in the four dimensions, It is a tensor value defined in the four dimensions. It is a time like tensor.

2. Proper Time ($\tau*$): It is a tensor quantity which depends on the trajectory path or exactly the curve that the particle follows. It is time with respect to the particle, in the particle's world.

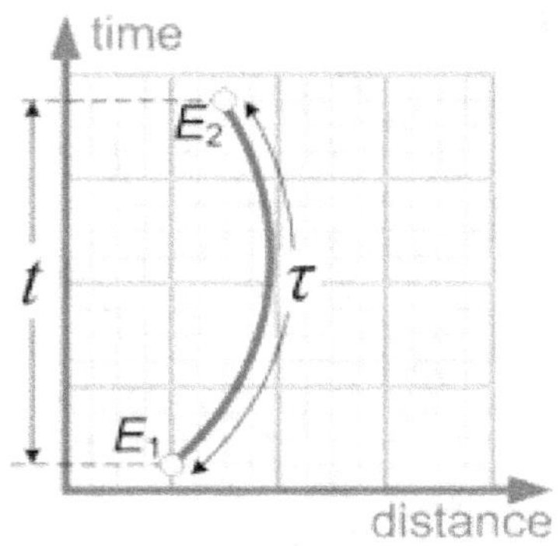

Figure 12: Proper time verses time

3. E*= (E/m)=Energy per unit rest mass (m)= 0.97(as set by software based on the initial velocity and the distance from the KBH)

4. Angular momentum of particle per unit rest mass (L/m)= 4000M= 4000(50M_{solar})=20,000 M_{solar}.

$$L = m(v \times r) = m(v_{ring}r)$$

$$(3.30)$$

$$\frac{L}{m} = v_{ring}r = 20,000\ M_{solar}$$

$$where\ r = \frac{20,000 M_{solar}}{0.196\ m/s} = 1020.4 M_{solar}\ km$$

5. Angular momentum per unit mass (J/M): It is the angular momentum of the Black hole. Magnitude has been taken as 0.5.
Angular Momentum $= I\omega^2$
I= Moment of Inertia about the FAOR
$\omega = Angular\ Velocity\ of\ rotation$ of Black Hole

$$a = \frac{I\omega^2}{M} = \frac{\frac{2}{5}MR^2\omega^2}{M} = \frac{2}{5}R^2\omega^2$$

$$(3.31)$$

$$\omega = 0.077\frac{rad}{s}$$
$R = 150km$ (for a 50M Black Hole)
$a = 0.5$

Note: To obtain the Moment of Inertia of the KBH, we have assumed it to be a solid sphere of ass $50M_{solar}$.

6. θ_{shell} : It is the initial angle which the four velocity vector makes with the radial vector form the Black Hole. Initial value has been taken as 90°(this simplifies the calculations for Angular Momentum.

7. $r^* = \dfrac{r}{M}$: Initial Radial distance per unit Mass of the black hole. It is a type of space measure or parameter for the potential at a point in the space around the black hole.

Initially $at\ t = 1\ second\ r^* = 20(approx.)$

From above calculations

$$r = 1020.4 M_{solar}\ km$$
$$\frac{r}{m} = \frac{1020.4}{50} = 20.408$$

(3.32)

Hence, verified with the software.

CHAPTER 4: SIMULATIONS AND RESULTS

With the help of the software out team has been able to visualize and understand various types of orbits around a Black Hole. The orbit around a Kerr Black Hole is governed by theories of special relativity, Tensor transformations, effect of space time curve, frame dragging, Gravitomagnetic moment. The space around a black hole is no ordinary space and can't be analysed by Newtonian Physics. The software equations are programed using Boyer-Lindquist Coordinates in Four Dimensions. The value of the Effective potential derived from the Kerr Matric helps to derive the orbit around the Black Hole. The Effective potential is the function of Carter's Constant which is the fourth conserved quantity for the particles orbit after Mass, Angular momentum and Energy.

Our Black Hole is 50 Solar Mass with Event Horizon being at a radius 150Kms. Our configurations have been taken at different spins of Black Hole.

Seven Configurations have been successfully simulated. Effects of Frame Dragging and Time Dilation have been observed through a detailed datasheet based on time, proper time, radial distance and elevation. The orbit around a KBH varies drastically with change in mass of black hole, spin, initial four velocity of particle. Simultaneously the graph of effective potential versus radial distance justifies the range within which the orbit is possible.

Each configuration has been explained in the following sections while the complete detailed input and output data has been included in Appendix B. The simulation pictures and explanations have been given in this section.

4.1 KBH Spinning Counter Clockwise

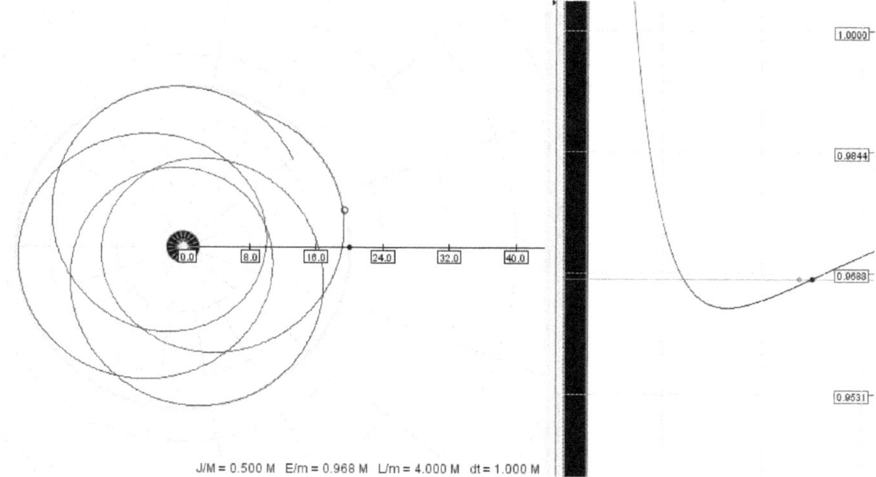

Figure 13: Trajectory at t=1858 sec

As we can notice in the data above, the particle is approaching the counter clockwise rotating Kerr black hole, the radius decreases. The particle is in a prograde orbit around the black hole. The effective potential changes as shown in the adjoining graph. The value of radial distances ranges within a limit beyond which the orbit is not stable.

4.2 KBH spinning clockwise. Particle Velocity vector is opposite to sense of rotation of Black Hole.

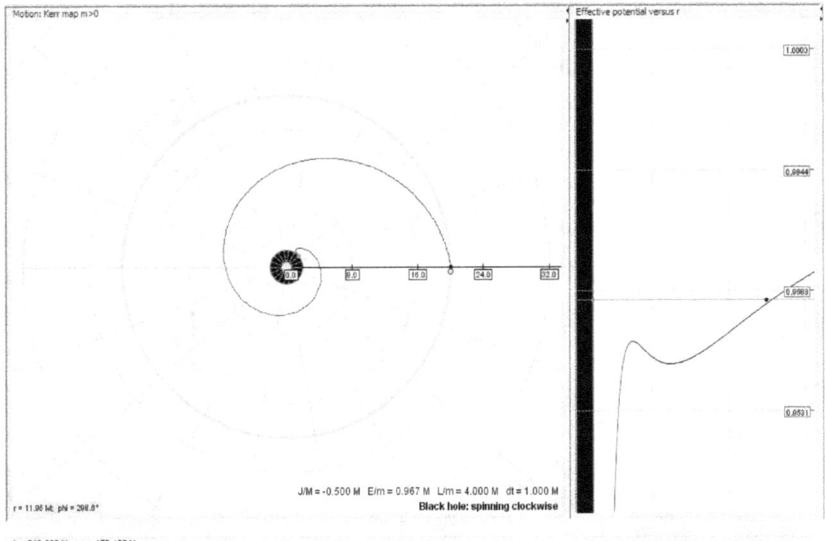

Figure 14: Trajectory at t=219 M

As shown in the figure above, the particle is approaching in a counter clockwise sense while the Kerr black hole is rotating clockwise. The particle is in a retrograde orbit around the black hole and ultimately succumbing to the event horizon. The particle takes a dragged orbit due to the effect of frame dragging and does not fall into it in a straight path.

4.3 Closet spinning orbit around a KBH (high time dilation).

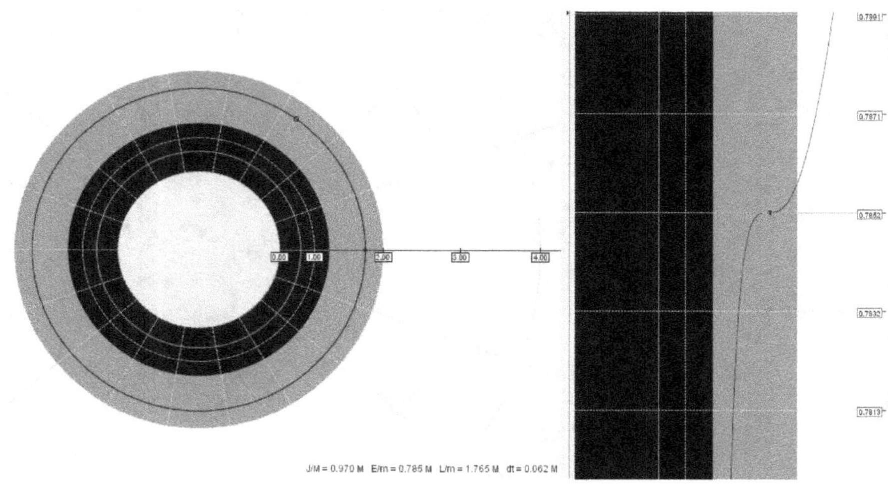

Figure 15: Trajectory of a particle in closest orbit possible

This is last stable circular orbit within the static limit of a rotating Kerr black hole. The beauty of this orbit is the effect of time dilation. As illustrated below. 23 seconds around the Black Hole is equivalent to 93 seconds of an outside observer.

t^*(sec)	r^*(km)	ϕ(radians)	τ^* (sec)
0	1.75751	0	0.0
93.3	1.75751	0	23.4501

As we can see from above simulated data, the radius of the orbit remains the same. The difference between the proper time and the time for an external observer has significant time dilation.

4.4 KBH spinning counter clockwise with four leaves orbit for sling shot manoeuvre

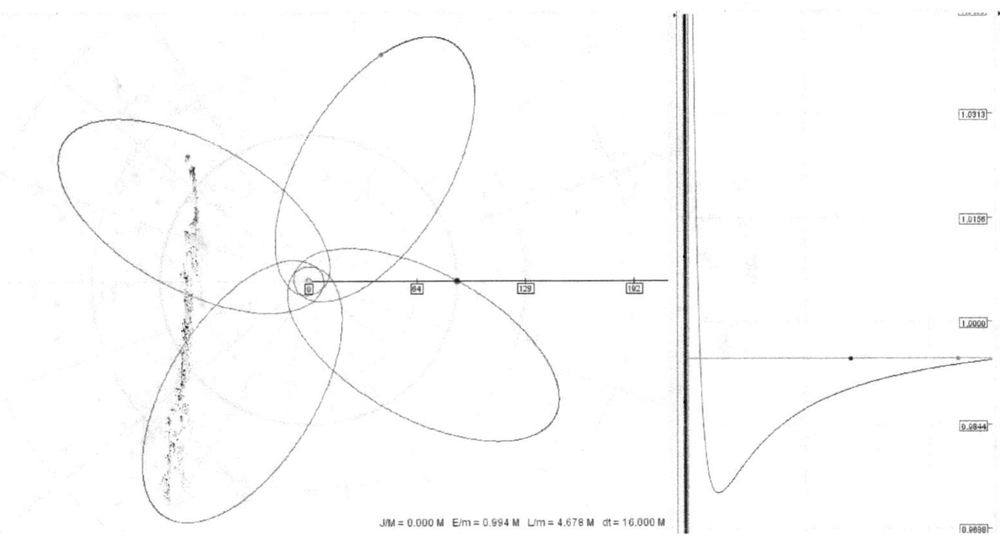

Figure 16: Trajectory for four leaf slingshot capable orbit

This orbit is called a four leaf orbit as there are four leaf-like paths being followed by the particle around our Kerr black hole. This kind of trajectory can be used for a slingshot mission or a mission which needs to observe the black hole from a closer point of view without much time dilation effect.

4.5 Orbit with nearly no time dilation

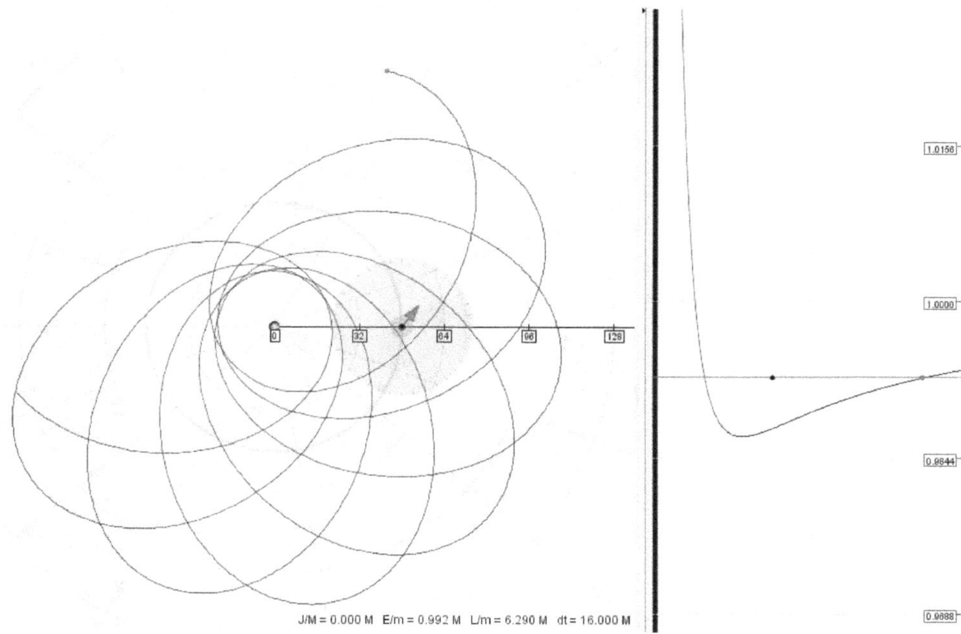

Figure 17: Trajectory of a particle with nearly no effective time dilation

This orbit is much like the four leaves orbit, but it gives us more opportunity to study the black hole from a closer point without much of time dilation. This orbit is suitable for near black hole studies and time consuming experiments.

t*(sec)	r*(km)	ϕ(radians)	τ* (sec)
320	74.47359	0.53329	311.88763
31	51.15733	0.07908	30.94443
16	49.63727	0.04073	15.46218
0	48.09481	0	0.0

4.6 KBH spinning counter clockwise with particle being trapped in the static limit

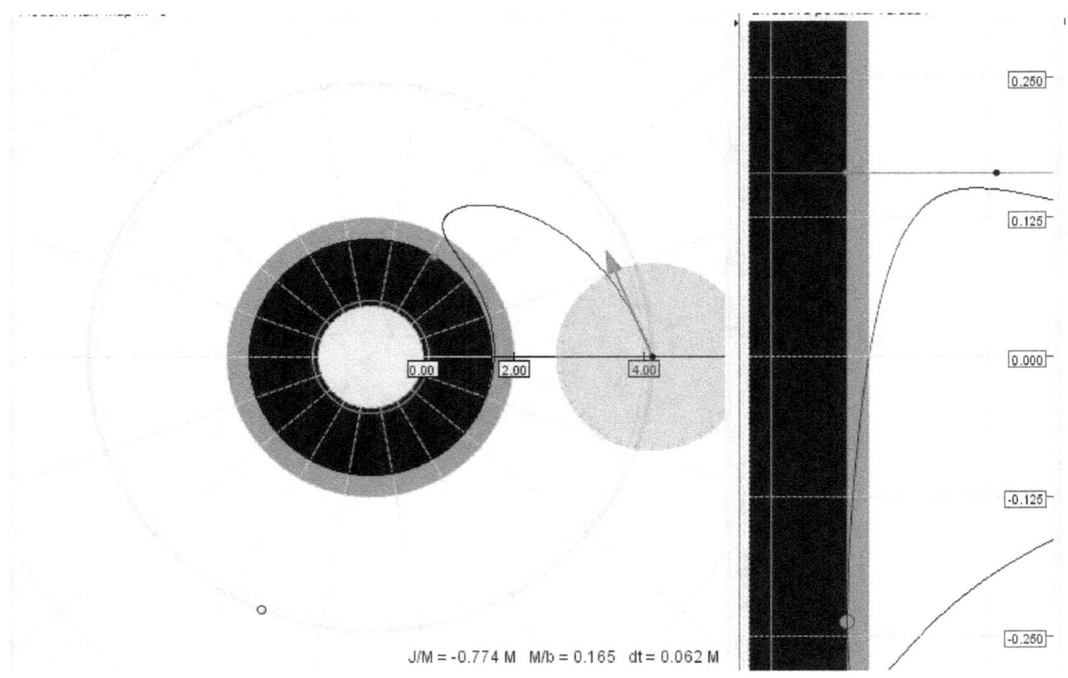

Figure 18: Particle being trapped in the static limit

This orbit shows a particle in a retrograde orbit around our Kerr black hole. At some point, the particle completely reverse its direction to come into a prograde orbit. The highlighted figures show the point where the particle reverses its direction and starts to fall into the event horizon. No proper time can be noted inside the static limit.

4.7 KBH with particle motion showing the static limit

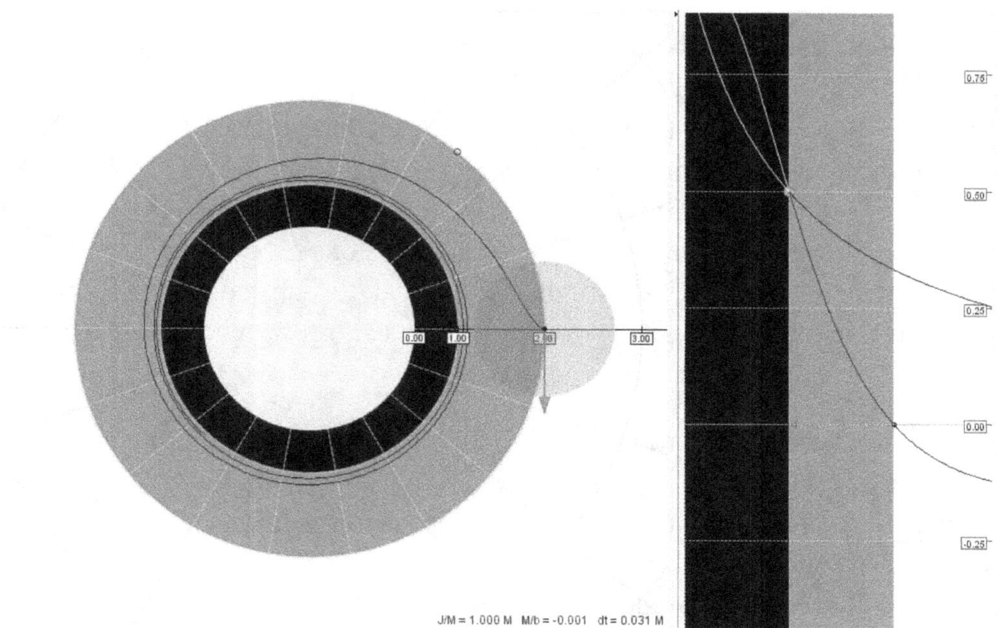

Figure 19: Orbit for Static Limit

The static limit, also known as the stationary limit, is the region around the black hole where nothing can remain stationary. The above data defines such an orbit in which the particle starts from outside the static limit and entering the region while continuously falling towards the event horizon. At some point the particle sets up an almost circular orbit around the event horizon.

CHAPTER 5: CONCLUSION

Black holes happen to be the place where quantum mechanics and gravity come together. Despite the fact that they are very big and very heavy, the structure of the way they contain information and the way they interact with the rest of the world, is highly quantum mechanical and so black hole seems to be the natural doorway, the place where we get really surprising paradoxes from, that we can explore in hopes to learn more about the connections between quantum mechanics and gravity. Our work here is thus to provide a substantial base for a future mission to a black hole so that we, as a race, can grow and quench our thirst for knowledge about our universe and our existence.

The space time curves near the region in proximity of the horizon. The frame dragging was successfully observed in the program and matched with the speculations. The orbit for a velocity in opposite direction to the sense of rotation of KBH eventually falls into the KBH whereas the orbit with a velocity in same sense as the sense of rotation gets trapped in a stable orbit around the event horizon at apoapsis and periapsis as recorded under r*.

JAVA based software has been used to simulate the results or an orbit around a KBH. Results have been obtained for prograde and retrograde orbits. Black hole of Mass 50 times the Solar Mass had been taken whose event horizon is 150km.Seven different configurations had been obtained around the KBH. Effects of Frame Dragging and Time Dilation were obtained very clearly showing relative changes in orbit plot at no spinning and spinning KBH. Leaf type orbits, molniya orbits and Last possible stable orbits have been observed. Our favourite reading includes a scenario in which we observed time dilation of upto 400 years, i.e. one year near our Black Hole was equal to outside clock's 400 years. This gives a result about the extent of hibernation that might have to be undertaken for humans to sustain life near a Kerr Black Hole.

When we look into the Universe today, we see that pretty much every large galaxy has a upper massive black hole in its heart. Even the Milky Way has a black hole at its core with a mass four million times that of the Sun.

REFERENCES

[1]   Thorne, K (1994). *Black Holes and Time Warps-Einstein's outrageous legacy*. W.W.Norton and Company.

[2]   Curtis, H.(2005).*Orbital Mechanics for Engineering Students.Burlington.*Elsevier Butterworth Heinemann.

[3]   C.W Mlsner, K.S Thorne and J A. Wheeler, *Gravitation.*(Freeman, San Francisco, 1973) p. 670

[1]   Komorowski,P.G, Valluri, S.R and Houde M.(2009).*A Study of Eliptical Last Stable Orbits About a Massive Kerr Black Hole.*

[2]   Frolov, V. P. and Novikov, I. D.(1997). *Black Hole Physics: Basic Concepts and New Developments.*

[3]   Aguirregabiria,J.M., Chamorro,A. and Virbhadra,K.S.(1996).*Energy and Angular momentum of charged rotating black holes.*

[4]   Stoghianidis, E. and Tsoubelis, D.(1986).*On The Periapsis Advance of Polar Orbits in the Kerr Family of Spacetimes.*

[5]   Johnston, M. and Ruffini, R.(1974).*Generalized Wilkins effect and selected orbits in a Kerr-Newman geometry*. Vol 10.

[6]   Tsoubelis,D., Economous, A. and Stoghianidis, E.(1987).*Local and Global Gravitomagnetic effects in Kerr spacetime.*.Vol. 36.

[7]   Kay, D.C. *Tensor Calculus. Schaum's Outlines*. McGraw Hill

[8]   Madore D.[Internet].[cited 2016 Oct 16].Available from:http://www.madore.org/~david/math/kerr.html.

[9]   Soberi,M .F. M. (2015).*Rotating Black Holes.*

[10]   Beyer,W. H., ed. (1991). *Standard Mathematical Tables and Formulae*, 29th Edition,CRC Press.

[11]   S. Chandrasekhar, *The mathematical theory of black holes.*(Oxford Umv. Press, Oxford, 1983) ch. 7.

[12]   R.A. van Patten and C.W.F Eventt, Phys Rev. Lett 36 (1976) 629

APPENDIX A

JAVA CODES

APPENDIX OUTLINE

A.1	INTRODUCTION	
A.2	JAVA CODE 1.1: Initial conditions for Boyer Lindquist L	
A.3	JAVA CODE 1.2: Initial Conditions Boyer Lindquist M	
A.4	JAVA CODE 1.3: Orbit Newton	

A.1 INTRODUCTION

JAVA has been used to simulate the orbit around a KBH which inputs the values a, E*, initial velocity, initial θ. The program observes the Boyer Lindquist coordinate system and follows the algorithm based on the Boyer Lindquist system of coordinates. Wherever necessary, Newtonian Physics has been used for simplicity. The Spin of the KBH in both directions has been analysed. The software needs conserved values to stay within constraints, these are m, M, a and l, After observing the Input values and analysing stored equations in light of those values, the software produces the desired trajectory and notes the values with t and τ. It gives radial distance and φ coordinates also. The frame dragging effect has been incorporated and frame dragging is observed by seeing the change in φ. Following program is the driving code of the work. Complete work involves working on a framed applet.

A.2 JAVA CODE 1.1: Initial conditions for Boyer Lindquist L

```
/*     */ package MajorProjects;

/*     */

/*     */ class InitialConditionsBoyerLindquistL extends InitialConditionsL

/*     */ {

/*     */   public InitialConditionsBoyerLindquistL(Orbit orbit, double a, double invB, double r, double sign, double dt)

/*     */   {

/* 117 */     super(orbit, a, invB, r, sign, dt);

/*     */   }

/*     */

/*     */   public InitialConditionsBoyerLindquistL(Orbit orbit, double a, double r, double theta0, double dt) {

/* 121 */     super(orbit, a, r, theta0, dt);

/*     */   }

/*     */

/*     */   public void adjustV0Theta0()

/*     */   {

/* 130 */     double invB = getInvB();

/* 131 */     double b = 1.0D / invB;

/* 132 */     double a = getA();

/* 133 */     double r = getR();

/* 134 */     double drdt = this.sign * Math.sqrt(Math.pow(a * a + r * r - 2.0D * r, 2.0D) * (r * r * r - 4.0D * a * b + b * b * (2.0D - r) + a * a * (2.0D + r)) / (r * Math.pow(r * r * r - 2.0D * a * b + a * a * (2.0D + r), 2.0D)));

/* 135 */     double dphidt = (2.0D * a - 2.0D * b + b * r) / (2.0D * a * a - 2.0D * a * b + a * a * r + r * r * r);
```

```java
/* 136 */    double v0SinTheta = (a * a + 2.0D * a * a / r + r * r) * (dphidt - 2.0D
* a / (r * (a * a + 2.0D * a * a / r + r * r))) / Math.sqrt(a * a - 2.0D * r + r * r);

/* 137 */    double x = 1.0D + 2.0D * a * a / (r * r * r) + a * a / (r * r);

/*     */    double sqrtx;

/*     */

/* 139 */    if (x < 0.0D)

/* 140 */      sqrtx = 0.0D;

/*     */    else {

/* 142 */      sqrtx = Math.sqrt(x);

/*     */    }

/* 144 */    double v0CosTheta = drdt * sqrtx / (1.0D + a * a / (r * r) - 2.0D / r);

/*     */

/* 146 */    double theta0 = Math.atan2(v0SinTheta, v0CosTheta);

/* 147 */    this.icData[3][1] = new Double(Math.toDegrees(theta0));

/*     */  }

/*     */

/*     */  public void adjustEmLmSign()

/*     */  {

/* 156 */    double theta0 = getTheta0();

/* 157 */    double PI = 3.141592653589793D;

/* 158 */    double epsTh = 0.001745329252D;

/* 159 */    if ((theta0 > PI - epsTh) && (theta0 <= PI)) {

/* 160 */      theta0 = PI - epsTh;

/*     */    }

/* 162 */    if ((theta0 < -PI + epsTh) && (theta0 >= -PI)) {
```

```java
/* 163 */      theta0 = -PI + epsTh;

/*     */    }

/* 165 */    if ((theta0 <epsTh) && (theta0 >= 0.0D)) {

/* 166 */      theta0 = epsTh;

/*     */    }

/* 168 */    if ((theta0 > -epsTh) && (theta0 <= 0.0D)) {

/* 169 */      theta0 = -epsTh;

/*     */    }

/* 171 */    this.icData[3][1] = new Double(Math.toDegrees(theta0));

/*     */

/* 173 */    double a = getA();

/* 174 */    double r = getR();

/* 175 */    double dphidt = Math.sin(theta0) * Math.sqrt(a * a + r * r - 2.0D * r) /
(r * r + a * a + 2.0D * a * a / r) + 2.0D * a / (r * (r * r + a * a + 2.0D * a * a / r));

/* 176 */    double invB = (2.0D * a * dphidt + r - 2.0D) / (dphidt * (r * r * r + a *
a * (2.0D + r)) - 2.0D * a);

/*     */

/* 178 */    if (Math.cos(theta0) > 0.0D)

/* 179 */      this.sign = 1.0D;

/*     */    else {

/* 181 */      this.sign = -1.0D;

/*     */    }

/* 183 */    this.icData[1][1] = new Double(1.0D / invB);

/*     */  }

/*     */
```

```java
/*    */  public void computeInitialState()

/*    */  {

/* 190 */    double eps = 1.0E-010D;

/* 191 */    double a = getA();

/* 192 */    double rHor = 1.0D + Math.sqrt(1.0D - a * a);

/* 193 */    double r = getR();

/* 194 */    double phi = 0.0D;

/* 195 */    double b = 1.0D / getInvB();

/*    */

/* 197 */    if (Math.abs(r - rHor) <eps) {

/* 198 */      if (r >rHor)

/* 199 */        r = rHor + eps;

/*    */      else {

/* 201 */        r = rHor - eps;

/*    */      }

/*    */    }

/*    */

/* 205 */    this.orbit.state[0] = r;

/* 206 */    this.orbit.state[1] = (this.sign * Math.sqrt(Math.pow(a * a + r * r -
2.0D * r, 2.0D) * (r * r * r - 4.0D * a * b + b * b * (2.0D - r) + a * a * (2.0D + r)) /
(r * Math.pow(r * r * r - 2.0D * a * b + a * a * (2.0D + r), 2.0D))));

/* 207 */    this.orbit.state[2] = phi;

/* 208 */    this.orbit.state[3] = 0.0D;

/* 209 */    this.orbit.state[4] = 0.0D;

/*    */  }
```

/*   */ }

/* Location: C:\Users\hp\Desktop\MajorProjects.jar

 * Qualified Name: MajorProjects.InitialConditionsBoyerLindquistL

A.3 JAVA CODE 1.2: Initial Conditions Boyer Lindquist M

```java
/*     */ package MajorProjects;

/*     */

/*     */ class InitialConditionsBoyerLindquistM extends InitialConditionsM

/*     */ {

/*     */   public InitialConditionsBoyerLindquistM(Orbit orbit, double a, double Em, double Lm, double r, double sign, double dt)

/*     */   {

/* 120 */     super(orbit, a, Em, Lm, r, sign, dt);

/*     */   }

/*     */

/*     */   public InitialConditionsBoyerLindquistM(Orbit orbit, double a, double r, double v0, double theta0, double dt) {

/* 124 */     super(orbit, a, r, v0, theta0, dt);

/*     */   }

/*     */

/*     */   public void adjustV0Theta0()

/*     */   {

/* 133 */     double Lm = getLm();

/* 134 */     double Em = getEm();

/* 135 */     double a = getA();

/* 136 */     double r = getR();

/* 137 */     double drdt = this.sign * (r * (a * a + r * (r - 2.0D)) * Math.sqrt((r * r * (2.0D - r + Em * Em * r) + Lm * Lm * (2.0D - r) - 4.0D * a * Em * Lm + a * a * (Em * Em * (2.0D + r) - r)) / (r * r * r))) / (Em * r * r * r - 2.0D * a * Lm + a * a * Em * (2.0D + r));
```

```java
/* 138 */    double dphidt = (Lm * (r - 2.0D) + 2.0D * a * Em) / (Em * r * r * r -
2.0D * a * Lm + a * a * Em * (2.0D + r));

/* 139 */    double v0 = Math.sqrt(drdt * drdt * r * (r * r * r + a * a * (2.0D + r)) /
((a * a + r * (r - 2.0D)) * (a * a + r * (r - 2.0D))) + (-2.0D * a + dphidt * r * r * r +
a * a * dphidt * (2.0D + r)) * (-2.0D * a + dphidt * r * r * r + a * a * dphidt *
(2.0D + r)) / (r * r * (a * a + r * (r - 2.0D))));

/* 140 */    double v0SinTheta = (a * a + 2.0D * a * a / r + r * r) * (dphidt - 2.0D
* a / (r * (a * a + 2.0D * a * a / r + r * r))) / Math.sqrt(a * a - 2.0D * r + r * r);

/* 141 */    double v0CosTheta = drdt * Math.sqrt(1.0D + 2.0D * a * a / (r * r * r)
+ a * a / (r * r)) / (1.0D + a * a / (r * r) - 2.0D / r);

/*    */

/* 143 */    double theta0 = Math.atan2(v0SinTheta, v0CosTheta);

/* 144 */    this.icData[4][1] = new Double(v0);

/* 145 */    this.icData[5][1] = new Double(Math.toDegrees(theta0));

/*    */  }

/*    */

/*    */  public void adjustEmLmSign()

/*    */  {

/* 154 */    double v0 = getV0();

/* 155 */    if (v0 >= 1.0D) {

/* 156 */      v0 = 0.99D;

/*    */    }

/* 158 */    if (v0 < 0.0D) {

/* 159 */      v0 = 0.0D;

/*    */    }

/*    */

/* 162 */    double theta0 = getTheta0();
```

```java
/* 163 */    double a = getA();

/* 164 */    double r = getR();

/* 165 */    double drdt = (1.0D - 2.0D / r + a * a / (r * r)) / Math.sqrt(1.0D + a * a
/ (r * r) + 2.0D * a * a / (r * r * r)) * v0 * Math.cos(theta0);

/* 166 */    double dphidt = v0 * Math.sin(theta0) * Math.sqrt(a * a + r * r - 2.0D
* r) / (r * r + a * a + 2.0D * a * a / r) + 2.0D * a / (r * (r * r + a * a + 2.0D * a * a /
r));

/* 167 */    double dtaudt = Math.sqrt(1.0D - 2.0D / r + 4.0D * a / r * dphidt -
drdt * drdt / (1.0D - 2.0D / r + a * a / (r * r)) - (1.0D + a * a / (r * r) + 2.0D * a * a
/ (r * r * r)) * r * r * dphidt * dphidt);

/* 168 */    double Em = (1.0D - 2.0D / r) / dtaudt + 2.0D * a / r * dphidt / dtaudt;

/* 169 */    double Lm = (1.0D + a * a / (r * r) + 2.0D * a * a / (r * r * r)) * r * r *
dphidt / dtaudt - 2.0D * a / r / dtaudt;

/*    */

/* 171 */    if (Math.cos(theta0) > 0.0D)

/* 172 */      this.sign = 1.0D;

/*    */    else {

/* 174 */      this.sign = -1.0D;

/*    */    }

/*    */

/* 177 */    this.icData[1][1] = new Double(Em);

/* 178 */    this.icData[2][1] = new Double(Lm);

/*    */  }

/*    */

/*    */  public void computeInitialState()

/*    */  {

/* 184 */    double eps = 1.0E-010D;
```

```java
/* 185 */    double a = getA();

/* 186 */    double rHor = 1.0D + Math.sqrt(1.0D - a * a);

/* 187 */    double r = getR();

/* 188 */    double Lm = getLm();

/* 189 */    double Em = getEm();

/* 190 */    double phi = 0.0D;

/*     */

/* 192 */    if (Math.abs(r - rHor) <eps) {

/* 193 */      if (r >rHor)

/* 194 */        r = rHor + eps;

/*     */      else {

/* 196 */        r = rHor - eps;

/*     */      }

/*     */    }

/*     */

/* 200 */    this.orbit.state[0] = r;

/* 201 */    this.orbit.state[1] = (this.sign * (r * (a * a + r * (r - 2.0D)) *
Math.sqrt((r * r * (2.0D - r + Em * Em * r) + Lm * Lm * (2.0D - r) - 4.0D * a *
Em * Lm + a * a * (Em * Em * (2.0D + r) - r)) / (r * r * r))) / (Em * r * r * r -
2.0D * a * Lm + a * a * Em * (2.0D + r)));

/* 202 */    this.orbit.state[2] = phi;

/* 203 */    this.orbit.state[3] = ((Lm * (r - 2.0D) + 2.0D * a * Em) / (Em * r * r *
r - 2.0D * a * Lm + a * a * Em * (2.0D + r)));

/* 204 */    this.orbit.state[4] = 0.0D;

/* 205 */    this.orbit.state[5] = 0.0D;

/*     */  }
```

```
/*    */ }


/* Location:          C:\Users\hp\Desktop\MajorProjects.jar
 * Qualified Name:    MajorProjects.InitialConditionsBoyerLindquistM
 * JD-Core Version:   0.6.2
```

A.4 JAVA CODE 1.3: JAVA CODE 1.6: Orbit Newton

```
/*     */ package MajorProjects;

/*     */

/*     */ import java.text.DecimalFormat;

/*     */ import org.opensourcephysics.tuleja.numerics.ODEAdaptiveSolver;

/*     */ import
org.opensourcephysics.tuleja.numerics.RK45MajorProjectsMultiStep;

/*     */

/*     */ public class OrbitNewton extends Orbit

/*     */ {

/*     */   public OrbitNewton()

/*     */   {

/* 10 */     this.twoPotentials = false;

/*     */   }

/*     */

/*     */   public void initializeVariables() {

/* 14 */     this.state = new double[] { 10.0D, 0.0D, 0.0D, 0.0D, 0.0D, 0.0D };

/*     */

/* 17 */     this.numPoints = 1500;

/* 18 */     this.ic = new InitialConditionsNewton(this, 0.0D, -0.05D, 4.0D,
20.0D, -1.0D, 1.0D);

/* 19 */     this.t = 0.0D;

/* 20 */     this.orbitData = new Double[this.numPoints][4];

/*     */   }
```

```java
/*    */
/*    */   public void initialize(double a, double r, double v0, double theta0, double dt, intnumPoints) {
/* 24 */     this.state = new double[] { 10.0D, 0.0D, 0.0D, 0.0D, 0.0D, 0.0D };
/*    */
/* 27 */     this.numPoints = numPoints;
/*    */
/* 29 */     this.ic = new InitialConditionsNewton(this, a, r, v0, theta0, dt);
/* 30 */     this.t = 0.0D;
/* 31 */     this.orbitData = new Double[numPoints][4];
/*    */
/* 34 */     this.odeSolver = new RK45MajorProjectsMultiStep(this);
/* 35 */     this.odeSolver.setTolerance(1.0E-006D);
/*    */
/* 38 */     this.odeSolver.initialize(this.ic.getDT());
/*    */
/* 41 */     reset();
/*    */   }
/*    */
/*    */   public void getRate(double[] state, double[] rates)
/*    */   {
/* 46 */     rates[0] = state[1];
/* 47 */     rates[1] = (state[0] * state[3] * state[3] - 1.0D / (state[0] * state[0]));
/* 48 */     rates[2] = state[3];
/* 49 */     rates[3] = (-2.0D / state[0] * state[1] * state[3]);
```

```java
/* 50 */    rates[4] = 1.0D;
/* 51 */    rates[5] = 1.0D;
/*    */  }
/*    */
/*    */  public double getR() {
/* 55 */    return this.state[0];
/*    */  }
/*    */
/*    */  public double getPhi() {
/* 59 */    return this.state[2];
/*    */  }
/*    */
/*    */  public double getTau() {
/* 63 */    return this.state[4];
/*    */  }
/*    */
/*    */  public double getT() {
/* 67 */    return this.state[5];
/*    */  }
/*    */
/*    */  public void setT(double t) {
/* 71 */    this.state[5] = t;
/*    */  }
/*    */
```

```java
/*    */   public double getRHorizon() {
/* 75 */     return 1.E-005D;
/*    */   }
/*    */
/*    */   public double getRInnerHorizon() {
/* 79 */     return 0.0D;
/*    */   }
/*    */
/*    */   public double rTorPlot(double r) {
/* 83 */     return r;
/*    */   }
/*    */
/*    */   public double rPlotTor(double rPlot) {
/* 87 */     return rPlot;
/*    */   }
/*    */
/*    */   public double getVmUpper(double r) {
/* 91 */     double Lm = this.ic.getLm();
/* 92 */     return -1.0D / r + Lm * Lm / (2.0D * r * r);
/*    */   }
/*    */
/*    */   public double getVmLower(double r) {
/* 96 */     double Lm = this.ic.getLm();
/* 97 */     return -1.0D / r + Lm * Lm / (2.0D * r * r);
```

```java
/*    */   }
/*    */
/*    */   public double getVmAtHorizon() {
/* 101 */     return getVmUpper(2.0D);
/*    */   }
/*    */
/*    */   public double getRingAngle() {
/* 105 */     return 0.0D;
/*    */   }
/*    */
/*    */   public String getOrbitInfo()
/*    */   {
/* 110 */     DecimalFormat format = new DecimalFormat("0.000");
/* 111 */     String sA = "J/M = ".concat(format.format(this.ic.getA())).concat(" M");
/* 112 */     String sEm = "E/m = ".concat(format.format(this.ic.getEm())).concat(" M");
/* 113 */     String sLm = "L/m = ".concat(format.format(this.ic.getLm())).concat(" M");
/* 114 */     String sDt = "dt = ".concat(format.format(getODESolver().getStepSize())).concat(" M");
/* 115 */     return sA + "   " + sEm + "   " + sLm + "   " + sDt;
/*    */   }
/*    */ }

/* Location:        C:\Users\hp\Desktop\MajorProjects.jar
 * Qualified Name:    MajorProjects.OrbitNewton
```

APPENDIX B

B.1	KBH spinning counter clockwise. Particle Velocity Vector is in same direction
B.2	KBH spinning clockwise. Particle Velocity vector is opposite to sense of rotation of Black Hole
B.3	Closet spinning orbit around a KBH (high time dilation)
B.4	KBH spinning counter clockwise with four petals orbit for sling shot manoeuvre
B.5	Orbit with effectively no time dilation
B.6	KBH spinning counter clockwise with particle being trapped in the event horizon
B.7	KBH with particle motion showing the static limit

B.1 KBH spinning counter clockwise. Particle Velocity Vector is in same direction

Inputs

$a^* \equiv a/M$	**0.5**
$E^* \equiv E/m$	0.968
$L^* \equiv L/(mM)$	4
$r^* \equiv r/M$ (initial)	20
v_{ring} (initial)	0.196
θ_{ring} (initial) degrees	90

L/m = 4000M

J/M = a = 0.5

Output:

t*(sec)	r*(km)	ϕ (radians)	τ* (sec)
1	20.00135	0.00943	0.930242
221	9.768146	3.800635	197.9337
222	9.758178	3.836262	198.7589
223	9.75003	3.871947	199.5839
224	9.743707	3.90768	200.4088
225	9.739215	3.943449	201.2335
226	9.736559	3.979241	202.0582
227	9.735739	4.015044	202.8828
228	9.736757	4.050847	203.7075
229	9.739612	4.086637	204.5322
230	9.744302	4.122402	205.3569
231	9.750822	4.158131	206.1818
232	9.759168	4.193811	207.0068
233	9.769332	4.22943	207.8321
234	9.781306	4.264978	208.6575
236	9.810643	4.335811	210.3092
237	9.827982	4.371074	211.1356
238	9.847084	4.406221	211.9623
239	9.867933	4.44124	212.7894
240	9.890512	4.476121	213.617
400	19.18056	7.505036	355.617

B.2 KBH spinning clockwise. Particle Velocity vector is opposite to sense of rotation of Black Hole

Inputs

$a^* \equiv a/M$	-0.5
$E^* \equiv E/m$	0.967
$L^* \equiv L/(mM)$	4
$r^* \equiv r/M$ (initial)	20
v_{ring} (initial)	0.199
θ_{ring} (initial) degrees	99.4

L/m = 4000M

J/M = a = 0.5

Output:

t*(sec)	r*(km)	φ (radians)	τ* (sec)
1	19.97043	0.009179	0.92971
206	3.503927	6.733824	169.8213
207	3.34726	6.840886	170.2192
208	3.182583	6.945884	170.5864
209	3.01137	7.046453	170.9189
210	2.836499	7.139144	171.2126
211	2.662652	7.219254	171.4639
212	2.496345	7.281016	171.6708
213	2.345093	7.318466	171.8337
214	2.215593	7.327004	171.9559
215	2.111648	7.304994	172.0435
216	2.033188	7.254298	172.1039
217	1.976997	7.179395	172.1443
219	1.912602	6.978627	172.1877
220	1.895759	6.862142	172.1986
221	1.884903	6.739424	172.2055
222	1.877967	6.612633	172.2099
223	1.873563	6.483219	172.2127
224	1.870776	6.352128	172.2144
225	1.869017	6.219974	172.2155
287	1.866025	-2.08336	172.2174

B.3 Closet spinning orbit around a KBH (high time dilation)

Inputs

$a^* \equiv a/M$	**0.97**
$E^* \equiv E/m$	0.785
$L^* \equiv L/(mM)$	1.765
$r^* \equiv r/M$ (initial)	1.758
v_{ring} (initial)	0.616
θ_{ring} (initial) degrees	90

Output:

t^*(sec)	r^*(km)	ϕ(radians)	τ^* (sec)
0	1.75751	0	0.0
5	1.75751	1.51518	1.25151
10	1.75751	3.03035	2.50301
15	1.75751	4.54553	3.75452
20	1.75751	6.0□071	5.00602
25	1.75751	7.57588	6.25753
30	1.75751	9.09106	7.50903
35	1.75751	10.6062	8.76054
40	1.75751	12.1214	10.012
45	1.75751	13.6366	11.2636
50	1.75751	15.1518	12.5151
55	1.75751	16.6669	13.7666
60	1.75751	18.1821	15.0181
65	1.75751	19.6973	16.2696
70	1.75751	21.2125	17.5211
75	1.75751	22.7277	18.7726
80	1.75751	24.2428	20.0241
85	1.75751	25.758	21.2756
90	1.75751	27.2732	22.5271
93.7	1.75751	28.3906	23.4501

B.4: KBH spinning counter clockwise with four leaves orbit for sling shot manoeuvre

Inputs

$a^* \equiv a/M$	**0.5**
$E^* \equiv E/m$	0.994
$L^* \equiv L/(mM)$	4.678
$r^* \equiv r/M$ (initial)	87.657
v_{ring} (initial)	0.109
θ_{ring} (initial) degrees	150.737

Output:

t*(sec)	r*(km)	φ(radians)	τ* (sec)
0	87.6574	0	0.0
16	86.165	0.00973	15.7198
32	84.6482	0.01981	31.4332
47	83.1062	0.03025	47.1396
63	81.5382	0.04109	62.8388
79	79.9435	0.05235	78.5303
95	78.3213	0.06407	94.2138
111	76.6707	0.07628	109.889
127	74.9907	0.08902	125.554
144	73.2803	0.10235	141.21
160	71.5386	0.11631	156.856
176	69.7643	0.13096	172.491
192	67.9562	0.14638	188.113
208	66.1131	0.16263	203.724
224	64.2334	0.17981	219.32
240	62.3158	0.19802	234.901
256	60.3586	0.21737	250.467
272	58.3601	0.23802	266.015
288	56.3184	0.26012	281.544
304	54.2314	0.28387	297.051
320	52.097	0.3095	312.536

B.5: Orbit with effectively no time dilation
Inputs

$a^* \equiv a/M$	**0.5**
$E^* \equiv E/m$	0.992
$L^* \equiv L/(mM)$	6.29
$r^* \equiv r/M$ (initial)	48.095
v_{ring} (initial)	0.164
θ_{ring} (initial) degrees	51.862

Output:

t*(sec)	r*(km)	ϕ(radians)	τ^* (sec)
320	74.47359	0.53329	311.88763
304	73.37276	0.51523	296.20174
287	72.25033	0.49663	280.52252
271	71.10605	0.47744	264.8503
255	69.93963	0.45763	249.18544
239	68.7508	0.43715	233.52836
223	67.53929	0.41596	217.87946
207	66.30483	0.39399	202.23921
191	65.04718	0.37118	186.6081
175	63.76609	0.34751	170.98667
160	62.46133	0.32285	155.3755
144	61.13271	0.29716	139.77521
128	59.78006	0.27033	124.1865
12	58.40326	0.24227	108.6101
96	57.00224	0.21287	93.04684
80	55.577	0.182	77.49761
64	54.12761	0.14952	61.96339
48	52.65428	0.11527	46.44527
31	51.15733	0.07908	30.94443
16	49.63727	0.04073	15.46218
0	48.09481	0	0.0

B.6: KBH spinning counter clockwise with particle being trapped in the static limit

Inputs

$a^* \equiv a/M$	**-0.774**
$b^* \equiv b/m$	6.066
$r^* \equiv r/M$ (initial)	4.122
θ_{ring} (initial) degrees	113.085

Output:

t*(sec)	r*(km)	φ(radians)	τ* (sec)
11.12499	2.11913	1.07531	0.0
11.06249	2.12796	1.07764	0.0
10.99999	2.13690	1.07980	0.0
10.93749	2.14593	1.08179	0.0
10.87499	2.15506	1.08361	0.0
10.81249	2.16429	1.08526	0.0
10.74999	2.17361	1.08674	0.0
10.68749	2.18303	1.08805	0.0
10.62499	2.19254	1.08919	0.0
10.56249	2.20214	1.09017	0.0
10.49999	2.21183	1.09097	0.0
10.43749	2.22161	1.09161	0.0
10.37499	2.23147	1.09208	0.0
10.31249	2.24142	1.09239	0.0
10.24999	2.25145	1.09254	0.0
10.18749	2.26157	1.09251	0.0
10.12499	2.27176	1.09233	0.0
10.06249	2.28203	1.09198	0.0
9.99999	2.29238	1.09148	0.0
9.93749	2.30280	1.09081	0.0
9.87499	2.31329	1.08998	0.0

B.7: KBH with particle motion showing the static limit

Inputs:

$a^* \equiv a/M$	1
$b^* \equiv b/m$	-866.707
$r^* \equiv r/M$ (initial)	2.005
θ_{ring} (initial) degrees	-90

Output:

t*(sec)	r*(km)	ϕ(radians)	τ* (sec)
28.78125	1.08845	9.77816	0.0
28.75000	1.08857	9.76392	0.0
28.71875	1.08868	9.74968	0.0
28.68750	1.08879	9.73544	0.0
28.65625	1.08890	9.72121	0.0
28.62500	1.08902	9.70697	0.0
28.59375	1.08913	9.69274	0.0
28.56250	1.08924	9.67851	0.0
28.53125	1.08936	9.66428	0.0
28.50000	1.08947	9.65005	0.0
28.46875	1.08959	9.63582	0.0
28.43750	1.08970	9.62160	0.0
28.40625	1.08982	9.60738	0.0
28.3750	1.08993	9.59316	0.0
28.34375	1.09005	9.57894	0.0
28.31250	1.09016	9.56472	0.0
28.28125	1.09028	9.55050	0.0
0.03125	2.00460	-3.5897	0.0

DIAGRAMS OF BLACKHOLE TRAJECTORIES

Direction of the black hole's spin (causing a frame dragging "force", which you could think of as a *torque* in the same direction as the black hole spins in)

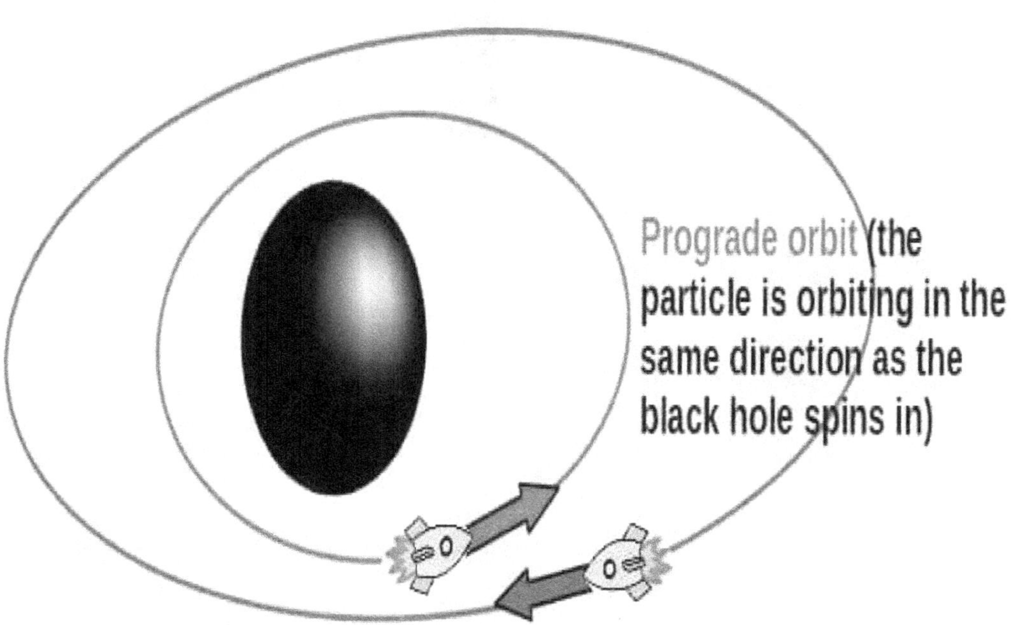

3D Kerr Black Hole Orbits

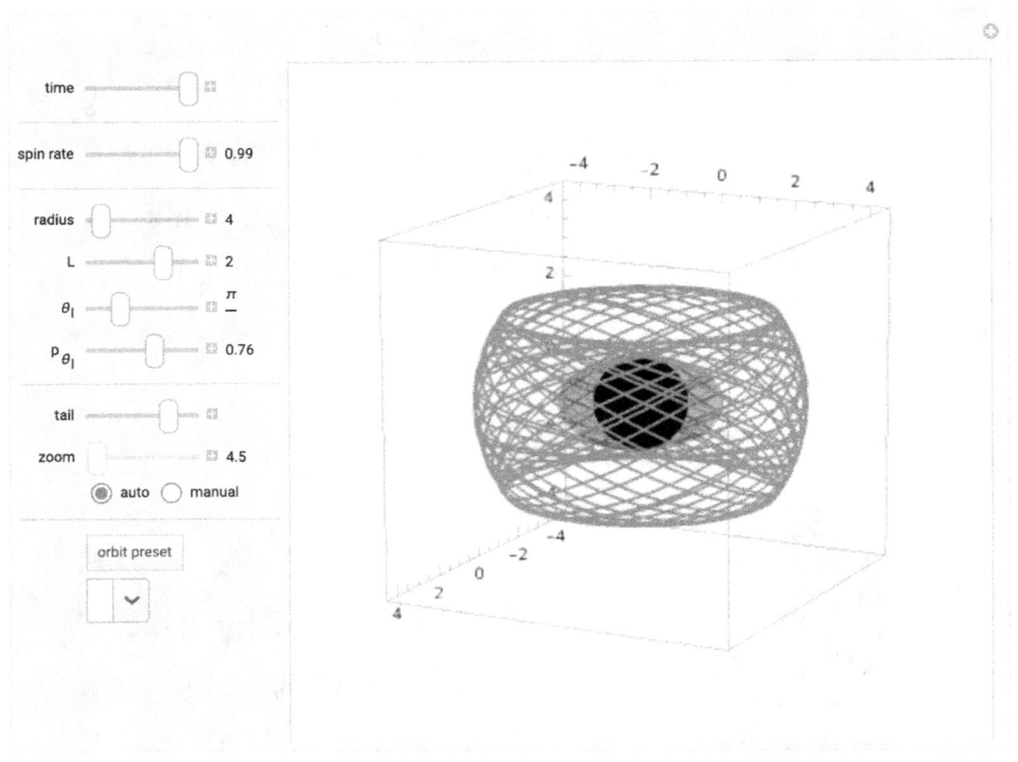

www.ingramcontent.com/pod-product-compliance
Lightning Source LLC
Chambersburg PA
CBHW081359160726
48000CB00010B/3408